Djallal Eddine Houari Adli
Khaled Kahloula
Kadda Hachem

Separation and fractionation techniques

Djallal Eddine Houari Adli
Khaled Kahloula
Kadda Hachem

Separation and fractionation techniques

Dialysis-filtration and centrifugation

ScienciaScripts

Imprint

Cover image: www.ingimage.com

This book is a translation from the original published under ISBN 978-613-8-46198-2.

Publisher:
Sciencia Scripts
is a trademark of
Dodo Books Indian Ocean Ltd. and OmniScriptum S.R.L publishing group

120 High Road, East Finchley, London, N2 9ED, United Kingdom
Str. Armeneasca 28/1, office 1, Chisinau MD-2012, Republic of Moldova, Europe
Printed at: see last page
ISBN: 978-620-4-07111-4

Table of materials

1. DIALYSIS

Dialysis is a technique for separating substances using their respective ability to pass through the pores of a membrane called a dialysis membrane.

1.1. General principles

The dialysis membranes usually used in biochemistry are in the form of elongated cylinders which must be closed at both ends and which contain the liquid to be dialysed. This cylinder is then called a dialysis "coil". It is placed in a container containing the liquid against which the dialysis is carried out or counter-dialysis liquid. (Figure 01)

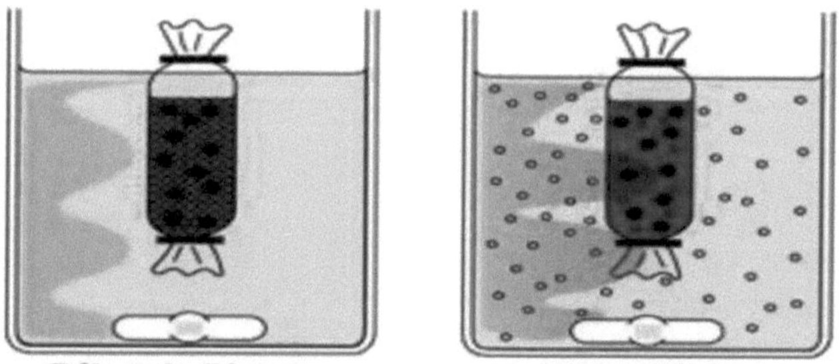

Figure 01 : General diagram of a dialysis

1.2. The membranes

Extemporaneously prepared collodion membranes are used and, at present, mainly cellophane tubes. The most commonly used are Visking membranes with a pore size of 2.4 nm (they allow the passage of pancreatic ribonuclease with an average diameter of 1.9 nm (molecular weight 25,000 daltons). Depending on the flat width of the cylinders, a distinction is made between:

Reference number	8/32	20/32	27/32	36/32	$1^{7/8}$ SS	$3^{1/4}$ SS
Width flat in cm	1	2,5	3,3	4,4	7,3 à 8,0	11,8 à 13,0

The porosity of these membranes can be modified by stretching in the longitudinal direction or by chemical treatment (acetylation or the action of zinc chloride at high concentration).

1.3. Factors influencing dialysis

A dialysis membrane should not be electrically charged in order to minimize the parasitic phenomena that could occur (repulsion by identical charge or adsorption on the membrane rises with the increase in temperature which amplifies the agitation of the molecules and thus increases the probability for a given molecule to cross the membrane. An important factor, which can considerably modify the dialysis rate, is the ratio of membrane surface to volume of the solution to be dialyzed.

For a given volume of solution to be dialyzed, the speed increases with the surface of the membrane (see below the application of dialysis). Dialysis also depends on the pH of the liquid to be dialysed and especially on the counter-dialysis pH: there is no precise rule because depending on the nature of the molecules, dialysable or not, the appearance of electrical charges on the molecule may or may not slow down the speed of dialysis.

Many substances can modify the speed of dialysis: thus a polypeptide molecule, adrenocorticotropin (or A.C.T.H), a hormone of the anterior pituitary gland, dialyzes against distilled water. If the water is replaced by a 0.15 M ammonium acetate solution, dialysis is practically inhibited. Often these changes in dialysis rate are due to the appearance of non-dialysable aggregates. This can be determined by studying the dialysis rate curve (Figure 02). The aspect schematized in A indicates the absence of aggregates, in B and C the break of the line indicates the presence of several molecular forms of the same substance (it should be noted that the aspect of a dialysis curve is that of an exponential and that it is necessary to use a semi-logarithmic representation to linearize the phenomenon). The addition of certain substances such as urea can modify the dialysis speed of many substances (generally in the direction of an increase); in fact urea acts on protein molecules by dissociating spatial structures which decreases the apparent radius of the molecule.

The dialysance of a molecule under well-defined conditions of membrane surface, pH, temperature, ionic concentration, etc., is often characterized by the half-dialysance time: time after which the concentration of the molecule studied has decreased by half in the dialysis coil (Figure 03) increases by the passage of water from the counter-dialysis to the coil. This phenomenon is due to the oncotic pressure of the proteins (the binding of mineral ions by the proteins and the balance of electrical charges on both sides of the membrane create a higher osmotic pressure in the compartment containing the proteins). In establishing a dialysis rate curve, these volume changes must be measured to correct the raw results.

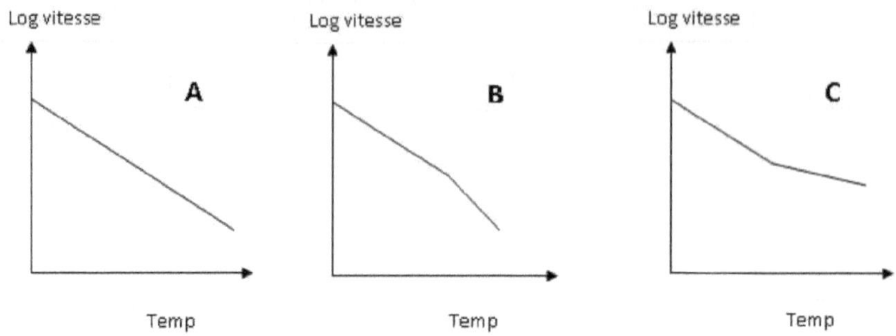

Figure 02: Study of dialysis as a function of time

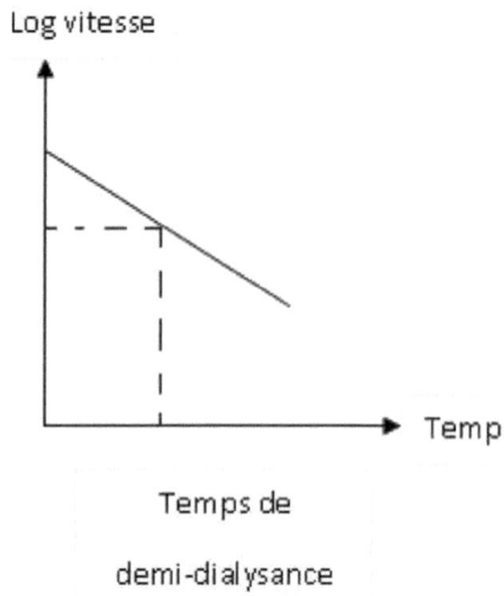

Figure 03: Measurement of the half-dialysis time

1.4. Dialysis techniques

1.4.1. Preparation of dialysis membranes

Dialysis membranes contain glycerine, sulphur compounds and heavy metals that must be removed by soaking. The usual technique consists of placing the membranes successively for one hour in a 10mM sodium bicarbonate solution, one hour in a dilute solution of EDTA (ethylene diamine tetra acetate: agent fixing by chelation of heavy metals), then in distilled water. The prepared membrane can be stored at 4°C for two to three days in water. If the storage time is longer, a bacterial growth inhibitor (sodium azide solution) must be added; if the tube dries out, it becomes unusable. Before use, the membrane must be rinsed with the solvent used for dialysis.

1.4.2. Preparation of dialysis pellets

The strings must be closed at both ends with a single knot with a return knot, which is closed with a second knot formed by the same wire used to tighten the first knot. In order to completely immerse the coil in the counter-dialysis liquid, it must be weighted with glass beads placed in the return (Figure 04). This will prevent it from floating on the surface of the counterdialysis fluid, as this would restrict the contact area and reduce the dialysis rate.

Proper agitation of the backflow fluid must be ensured to avoid the formation of concentration gradients of diffusible substances in the beaker that usually contains the backflow fluid. This is particularly useful when the volume of the counterdialysis fluid is large. This fluid must be renewed frequently in order to accelerate the disappearance of diffusible substances in the coil.

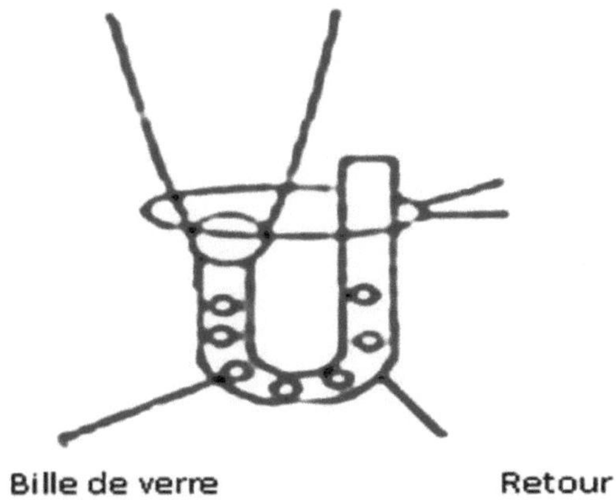

Figure 04: Closing the dialysis tubing

1.4.3. Special assemblies

1.4.3.1. Circular Dialysis

The described set-up allows a fast dialysis (flow rate of the counter-dialysis liquid: 2 ml /minute) the counter-dialysis liquid flows along the coil. The chamber must be saturated with water vapour in order to prevent evaporation phenomena from concentrating the buffer solution used for counter-dialysis (Figure 05).

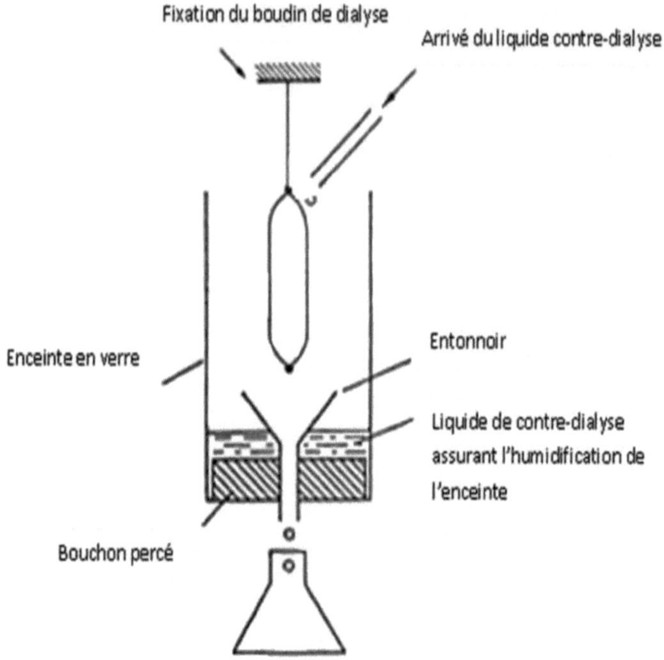

Figure 05: Circular Dialysis

1.4.3.2. High volume dialysis

It is then interesting to use a continuous flow system (for example, a dialyzer used in automatic analysis). The liquid to be dialysed passes in contact with the counter-dialysis liquid, from which it is separated only by a membrane: dialysis takes place over a large surface area. The counter-dialysis liquid is constantly renewed, while the dialysis liquid is constantly returned to the system by a peristaltic pump (Figure 06).

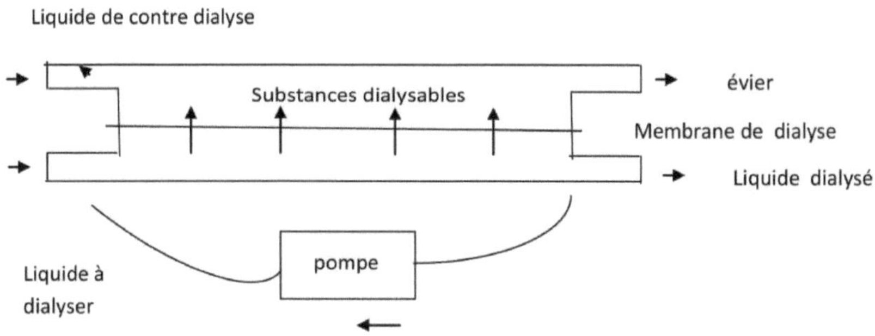

Figure 06: Dialysis in continuous flow

1.4.3.3. Dialysis on small volumes

In order to increase the dialysis surface area, a glass tube weighted with mercury and closed at both ends is inserted into the coil. The tube occupies most of the volume of the coil and the liquid to be dialyzed is thus in contact over a larger surface area with the counter-dialysis liquid via the membrane (Figure 07).

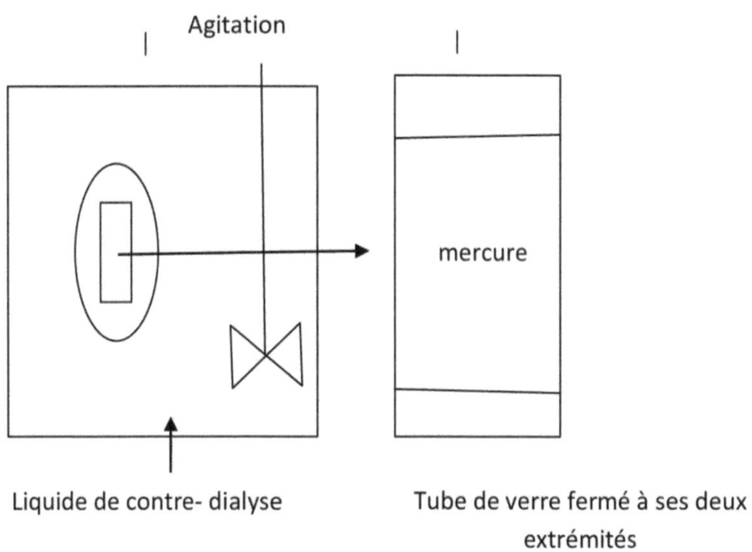

Figure 07: Dialysis on small volumes

1.4.3.4. Constant volume dialysis

The described set-up allows a positive pressure to be applied to the fluid to be dialyzed. This pressure can compensate for the oncotic pressure of the proteins and thus prevents the volume increase normally observed in this case (Figure 08).

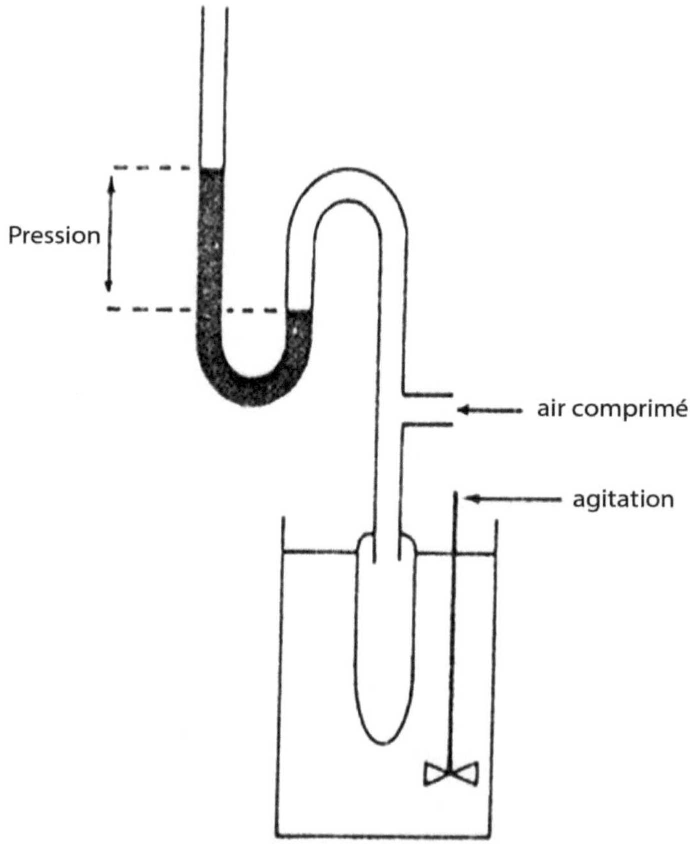

Figure 08: Constant volume dialysis

1.4.3.4. Electrodialysis

A separation technique similar to dialysis but using an electric field to enhance ion transport across the membrane. The establishment of a current of a few mA allows the passage of anions to the cathode and cations to the anode. Mineral ions are thus removed without loss of organic products.

The operation takes place in a three-compartment tank (Figure 09). The central compartment, which contains the liquid to be treated, is separated by a hemi-permeable membrane from the two lateral compartments, one containing an anode (anodic medium) and the other a cathode (cathodic medium). Thus, the anions and cations attracted by the corresponding electrodes are eliminated and concentrated outside the medium to be treated. A voltage of a few volts, usually 2 volts, is sufficient.

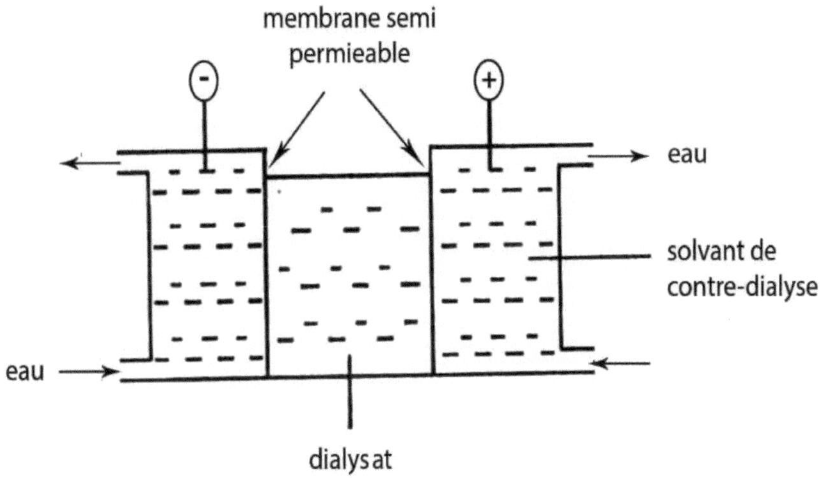

Figure 09: Principle of electrodialysis.

Electrodialysis is used for demineralization and deacidification in the dairy industry. The thermal rise produced by the current does not allow this use with thermolabile products.

1.4.3.4. A. Osmosis

Osmosis is a physical phenomenon in which a flow of water occurs between two solutions of different molecular concentrations separated by a hemi-permeable membrane, in such a way that the aqueous solvent flows from the less concentrated solution to the more concentrated one until the pressure exerted on the more concentrated solution reaches a value called osmotic pressure (Po), the solute(s) cannot change compartment.

Concentration can also be achieved by osmosis if the membrane is impermeable to solutes. This is the case of dialysis against a macromolecular solution such as polyvinylpyrrolidone (PVP) or polyethylene glycol (PEG) solutions. Commercially available PEGs with average molecular weights of 30,000 are preferred, which limits their diffusion in the liquid to be concentrated. The quality of liquid to be concentrated is placed in the cellulose tube which is dialysed for a period varying from 4 to 24 hours, depending on the desired concentration coefficient in a 60% PEG solution. The concentration is generally done at 4°C. By this means, the passage of PEG in the concentrate is limited to the maximum. It is essential to carefully record the amount of liquid before and after concentration to establish the concentration factor.

1.4.3.4. B. Reverse Osmosis

A very fine separation technique based on the transfer of water between two solutions of different concentrations separated by a semi-permeable membrane, which is done "in reverse" due to the effect of an artificial osmotic pressure (30 to 80 arms) that counteracts the natural osmotic flow from the more concentrated solution to the less concentrated one, leaving low molecular weight contaminating particles (dissolved solids, bacteria, etc.) in its wake (Figure 10). The higher the concentration and the lower the molar mass, the greater the osmotic pressure. The concentration takes place without temperature increase. The low flow rate of this method can be compensated for by the use of large surface filters.

The membranes used have a cut-off between 150 and 1000Da. Two types of membranes are used:

- Cellulose membranes: very permeable and selective, but sensitive to chemical hydrolysis. They are only used at temperatures below 40°C and in a pH range of 3 to 8. They can be used up to about 60 bar.

- Membranes made of organic polymers, mainly polysulfones, which are resistant to extreme pH and temperatures of 70°C. The flow is limited to 150 l/h/m2. Polyamide membranes are also used, which remove 90 to 98% of inorganic elements, non-ionic elements and organic molecules with a molecular weight greater than 100 Da.

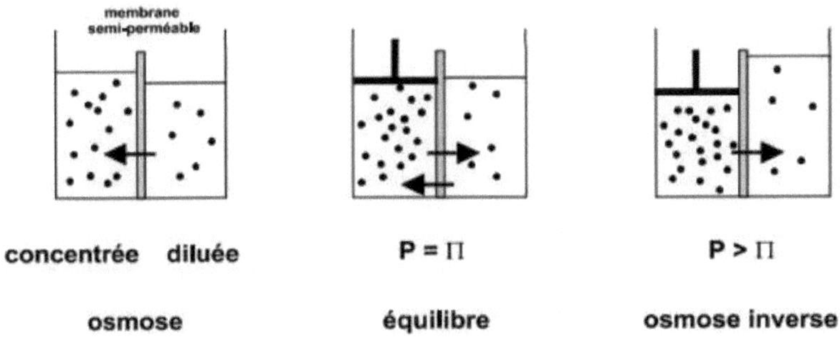

Figure 10: Principle of reverse osmosis.

1.5. Application of dialysis

Electrodialysis is used to desalinate samples before performing chromatography on paper. The dialysis by coil allows for example the following operations:

* Remove diffusible products, e.g. $(NH4)_{2SO4}$, from solutions.

* Equilibrate a protein solution with a buffer that changes the pH of the protein solution.

* Conversely, to bring in a protein solution a diffusible substance placed in a dialysis coil: it is thus possible to bring progressively molecules of $(NH4)_{2SO4}$ in a protein solution.

* Concentrate a protein solution either by placing the dialysis rod containing the solution in a stream of air or by placing the rod in powdered polyethylene glycol. The water leaves the dialysis rod to solubilize the polyethylene glycol, which is non-diffusible and therefore does not penetrate the protein solution.

* Desalination of brackish water.

* In the dairy industry: concentration of whey, skimmed milk and whole milk,

* Concentration of active substances (antibiotics, amino acids, vitamins, etc.),

* Purification and concentration of natural aromas,

* Concentration of fruit juices,

* Medical practice.

2. FILRATION

There are two types of filters: "depth" and "screen" filters.

The "deep" filters (Figure 11) are made of fibrous substances (cellulose magnet, cotton, coal, etc.). In both cases, the particles that are to be stopped by the filtration will be blocked in a labyrinth of tortuous channels existing inside the filter itself. The efficiency of a depth filter therefore increases with the thickness of the filter; on the other hand, it decreases when the pressure applied to the filter increases. Efficiency can be defined as the percentage of particles retained by a filter. Screen" filters (Figure 12) retain all particles larger than the pore size on their surface. Increasing the thickness of the filter is obviously of no interest, except to increase its mechanical strength.

Increasing the pressure on the filter accelerates the flow rate without decreasing the filtration efficiency. The only parameter that affects the efficiency is the pore size. However, the retention of all particles on one plane leads to a rapid clogging of these filters.

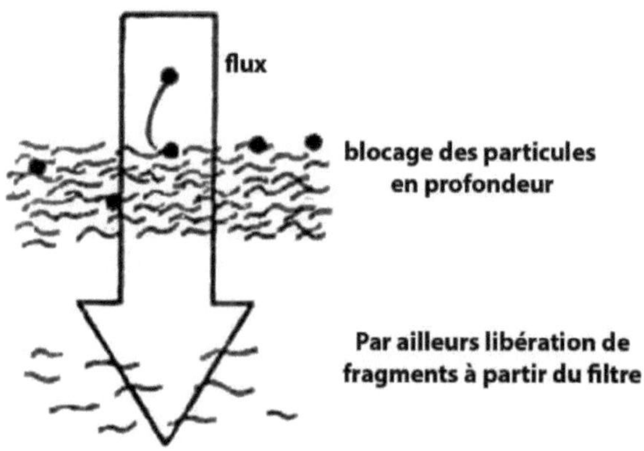

Figure 11: Depth Filters

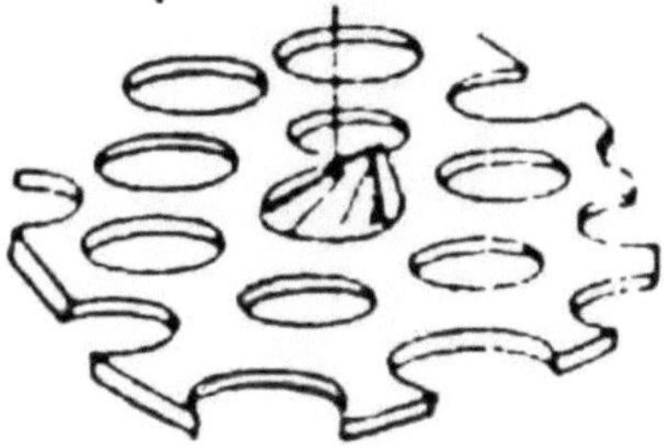

Figure 12: Screen filter

2.1. Deep" filters

2.1.1. Classic filter paper

It is used in the form of filters without folds to collect a precipitate, or in the form of pleated filters which, for the same size, have a larger filtration surface than the filters without folds and are therefore used for the filtration of large volumes of liquid. There are several qualities of paper and, in practical terms, a distinction must be made between filters for rapid filtration (Durieux red band or No. 4 and No. 90 Watman), filters for medium-speed filtration (Durieux white band or No. 1 and No. 7 Watman), and filters for very slow filtration (Durieux blue band or No. 5 and No. 6 Watman).

All these filters are used on a support, usually a conical funnel or a buchner.

2.1.2. Fiberglass filters

They are always used flat on a funnel with a filtered plate. They have a circular shape with a small diameter and must always be used unfolded, because folding weakens them and cracks them.

The pores of these filters are narrower than those of filter papers and they are usually used to remove particles smaller than 1ùm.

These filters can be preheated to 500°C to remove all traces of organic contaminants before gravimetric use. Their high filtration rate allows them to be used under a moderate vacuum (Figure 13).

These filters are sometimes used to collect radioactive precipitates and after drying can be directly immersed in scintillating mixture.

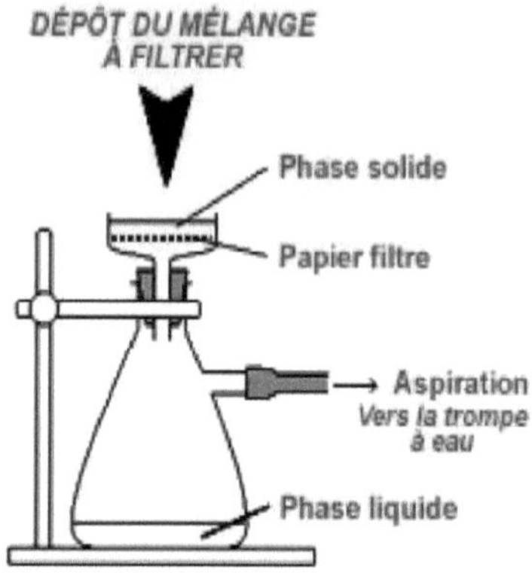

Figure 13: Buchner filtration

2.1.3. Filter tubes

They are made of glass microfibers bonded together with a resin. The filter tubes are in the form of hollow cylinders that can be mounted on connecting pieces. They allow the rapid filtration of liquid or even air. The retention is about 2ùm. These tubes are used in biochemistry to provide automatic analysis of reagents free of any contaminating particles, but also to filter gases or to quickly clarify large volumes of turbid solutions (Figure 14).

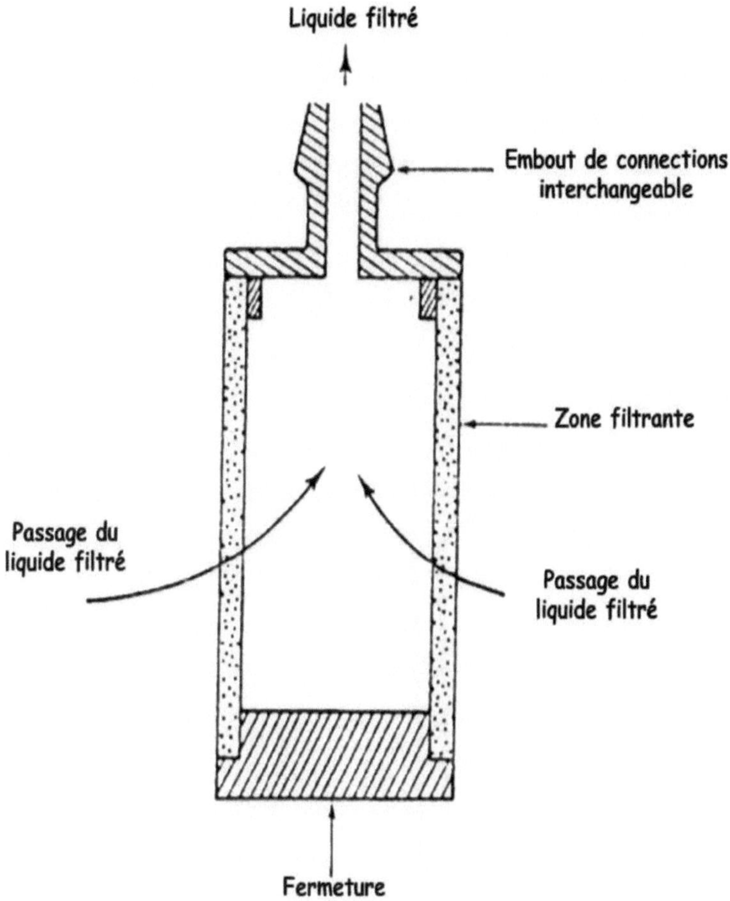

Figure 14: Mounted filter tube

2.1.4. Phase separating filter paper

The Watman I P S phase separator filter paper can replace a separating funnel. It is indeed a hydrophobic paper which allows to separate easily and quickly aqueous solutions from solvents immiscible with water. Simply pour both phases simultaneously onto the filter paper placed in a glass funnel: the aqueous phase remains on the filter while the organic phase passes through it.

The aqueous phase should not be left on the filter for too long after separation as the water will eventually pass through the filter. These filters should never be used under vacuum, even if the vacuum is moderate. Finally, they should not be used if the aqueous phase contains a high concentration of a polar solvent (e.g. ethanol) which lowers its surface tension.

2.1.5. Sintered plates

These are porous plates made of agglomerated glass powder. They exist in six pore sizes (Table 01) and are widely used for vacuum filtration. After use, the sintered plates must be cleaned with the sulfochromic mixture, then washed thoroughly with distilled water. (Figure 15)

Figure 15: Filter plate funnel

Table 01: Sintered plates

Porosity	Pore diameter in micrometers	Applications
00	200-500	Special uses
0	150-200	To stop large particles
I	90- 150	Filtration of coarse precipitates; gas scrubbing
2	40- 90	for medium-sized precipitates; gas filtration
3	20- 40	For fine precipitates, mercury filtration
4	10- 20	For ultra-fine precipitates

2.1.6. Screen" filters

The most commonly used are the Millipore filter membranes. They are made of cellulose polymers with a very large number of rigorously calibrated pores. The dimensions of these pores vary according to the filters from 0.025 p.m to 8 p.m while the thickness of the filter is 0.15 mm. These filters resist for at least 30 minutes both to temperatures of 125°C and to high pressures, which allows their sterilization by autoclaving.

The standard quality or M.F. filter is not resistant to certain solvents (ethanol, methanol, acetone, etc.) but the solvinert series is very resistant to these organic solvents; the same applies to Teflon filters.

Screen filters stop particles at their surface so that their dirt-holding capacity is low. For this reason, screen filtration is usually preceded by filtration through an asbestos-cellulose or glass fibre depth filter (often referred to as a pre-filter).

Filtration is almost always carried out under pressure (very rarely under vacuum) of the order of 1.5 bar, the limit of resistance of the pre-filters, the filters themselves being able to withstand much higher pressures.

The filter-prefilter assembly is enclosed in a tightly sealed metal or plastic case. The metal boxes containing the filter and pre-filter can be sterilized in an autoclave (wrapped in kraft paper and never in aluminium foil which does not allow the sterilizing steam to enter).

To prevent the clogging of small pore filters, another filter with larger pores must be inserted between the prefilter and the filter; these two filters must be separated by dacron cloths to prevent clogging caused by the simple superimposition of filters (Figure 16). Up to four filters can be superimposed in this way.

Screen" filters are used for sterilizing filtrations: filters with a pore size of 0.22um must be used (when these sterile solutions are to be used for cell culture media, the filter must be rinsed thoroughly with distilled water in order to eliminate the triton X 100 contained in the filter, as this product is toxic for the cells).

Sterilizing filtration is one of the applications of choice for "screen" filters, but they are also used in cytology: the filtration of a biological medium makes it very easy to collect the cells it contains. The filter can then undergo a chemical treatment which colours its cells and direct observation under a microscope is possible because the oil used for immersion has a refractive index of 1.51 which "transparently" colours the filter.

In order to simplify assembly and sterilization problems, it is possible to use filters mounted in a single-use plastic case (Twin 90).

The most commonly used Millipore membranes are made of cellulose nicrate, which can adsorb proteins. The same is true of the membra-Fil (JoHns Manville) and sartorius membranes. These adsorptions are very weak with the Nucleopore Serlabo membranes and the Unipore Bio Rad membranes (made of polycarbonate).

Gellman membranes come in a variety of structures: Mitricel GA and TCM (cellulose triacetate), Metricel GN (a mixture of cellulose esters), Metricel Vinyl (made of polyvinyl chloride), Metricel DM and Acropor (a copolymer of acrylonitrile and PVC).

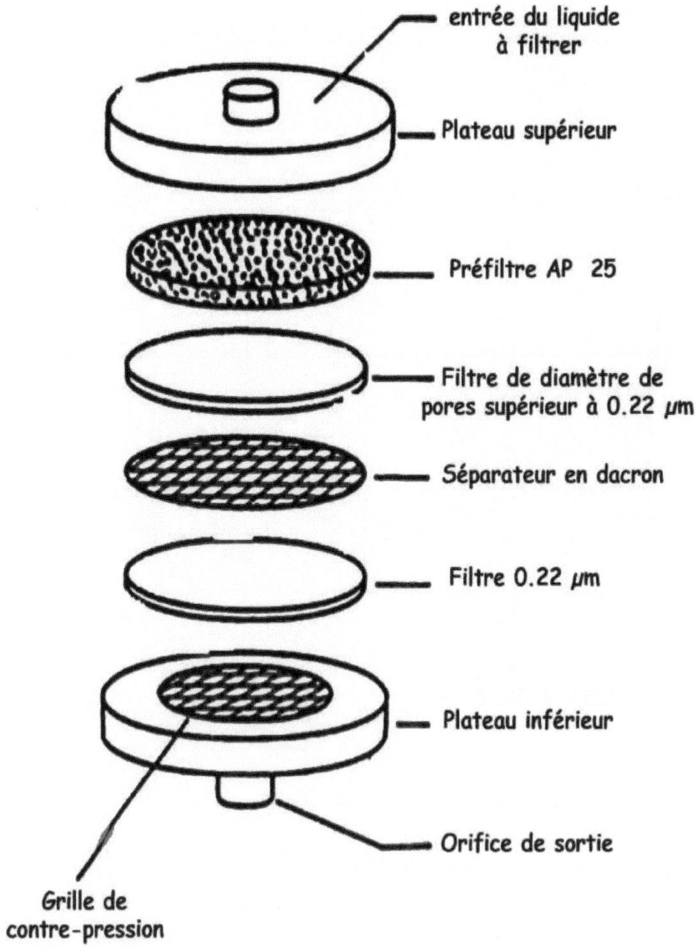

Figure 16: Example of a screen filter assembly

2.2. ULTRAFILTRATION

2.2.1. Concentration by ultrafiltration on membrane

Ultrafiltration is the process of separating under pressure macromolecules dissolved in a solvent, according to their size (from 0.001 to 0.25 um), through a membrane with extremely fine pores (diameter less than 0.05 thousandth of a millimetre). The liquid containing the molecules to be separated is placed in a compartment limited by a membrane that only allows the solvent and small molecules to pass through (ultra-filtrate) and retains the solutes of dimensions greater than the molecular retention threshold of the membrane (retentate) (Figure 17). This results in a progressive and selective concentration.

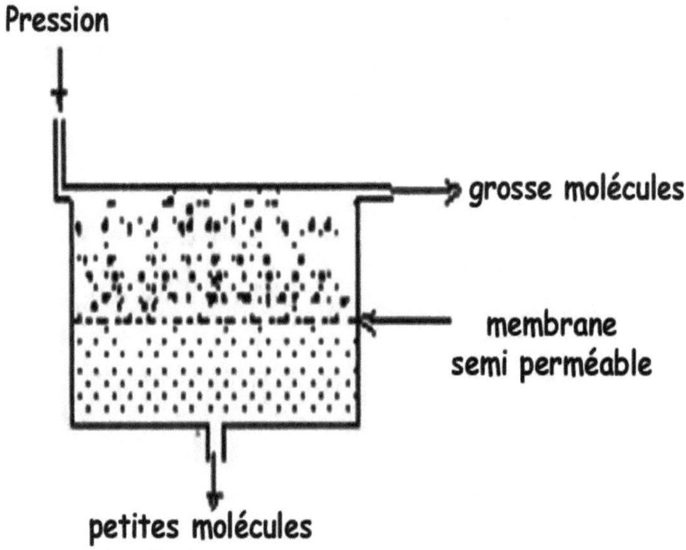

Figure 17: Ultrafiltration by pressurization.

The retention threshold of an ultrafiltration membrane is called the "Nominal Molecular Weight Limit" (NMWL), a term used to characterize a membrane. Currently available ultrafiltration membranes offer a selective choice of nominal molecular weight cut-offs from 500 to 1,000,000, i.e. a size of 10 to 100 1.1111. These membranes are produced in a wide variety of polymers suitable for particular applications: acids, bases, alcohols, high temperature, etc. Zirconium oxide membranes on a graphite support can be used at temperatures of around 3,000°C and pH values of 1 to 14.

The retention coefficient (R) of a membrane for any solute and for each membrane is determined by the formula :

$$R = \text{In}(C_f/C_0)/\text{Ln}(V_0/V_f)$$

Where C_f is the final concentration.

C_0 initial concentration

V_0 initial volume

V_f: final volume.

When the macromolecule is completely retained by the membrane, R=1 (100% retention); when the molecule passes freely, R=0.

The solution can be pressurized (Figure 17) either by pressurized nitrogen (8 to 10 bar) or by increasing the centrifugal force by centrifugation (Figure 18).

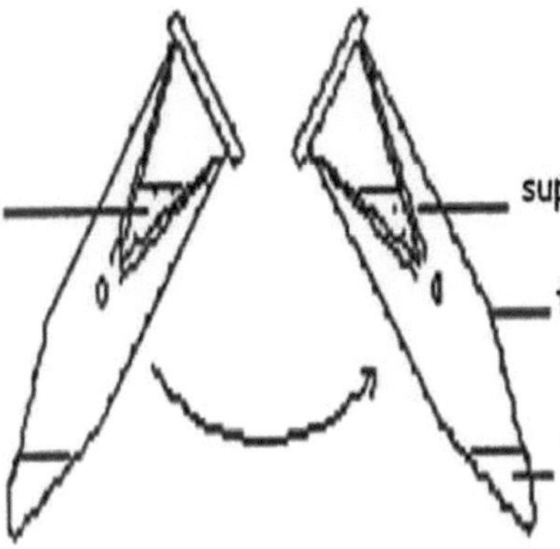

Figure 18: Ultrafiltration by centrifugation.

Application example:

- In the dairy industry: concentration of whey proteins, preparation of concentrated syrups from an enzymatic hydrolysate of lactose, preparation of cheese, etc.

- In biochemistry: exchange or removal of salts, sugars, animate acids, low molecular weight inhibiting substances or changes in ionic strength and pH; purification and concentration of enzymes; fractionation by separation of molecules of different sizes or shapes. Ultrapurification is widely used at the end of the purification process, as it allows the removal of eluting agents (in the case of a chromatographic step) and above all the concentration of protein solutions.

When we want to reach high protein contents, we proceed to a dia-filtration (addition of solvent to the solution during ultrafiltration) which allows to purify even more the macrosolutes of the solution by washing the retained species. This method is to be preferred to conventional dialysis which is often long and inefficient at low concentrations.

- In the food industry: concentration by removing solvent or water (e.g. fruit juice, molasses, grape must, etc.), concentration of egg white (from 12.3% to 25%);

- In the pharmaceutical industry: preparation of ultrapure waters.

2.2.2. Ultrafiltration on membrane

The most commonly used membranes are Diaflo membranes (Amicon). Table I shows the stopping levels of the different types of Diaflo membranes. These membranes are used in ultrafiltration cells (of which there are many models, see Figure 19). The force allowing ultrafiltration is a pressure exerted on the liquid to be filtered by nitrogen: these pressures are low and vary with the type of membrane used.

Diaflo membranes are coated with glycerine which must be removed before use by rinsing with water. They can be reused and are then kept wet. Ultrafiltration can be carried out by centrifugation: the driving force of ultrafiltration is then the centrifugal force. The liquid to be ultrafiltered (maximum volume 7ml) is placed in a cone formed by a glued membrane.

This cone is placed in a plastic support with a lower hole and itself placed on the upper end of a centrifuge tube.

The liquid remaining in the cone is thus concentrated into molecules with a molecular weight greater than 50,000 (Centriflo CFSOA) or 25,000 (Centriflo F25).

Membrane ultrafiltration can have many applications:

* Concentration of macromolecules (e.g. proteins);

* Removal of low molecular weight substances (below the membrane's stopping limit): this is the equivalent of rapid dialysis or diafiltration, which consists of widely diluting the liquid to be ultrafiltered before starting the ultrafiltration operation;

* Application of the protein concentration technique to column chromatography: this operation allows the removal of most of the solvent;

* Study of the binding between macro-molecules and low molecular weight molecules ;

* Separation of viruses using special membranes with stop limits of XM 300 (M.P.: 300,000) and DP02, DP05, DP06 for particle sizes between 100 and 1000 nm.

Several companies market membranes with different constitutions and properties (Table 02).

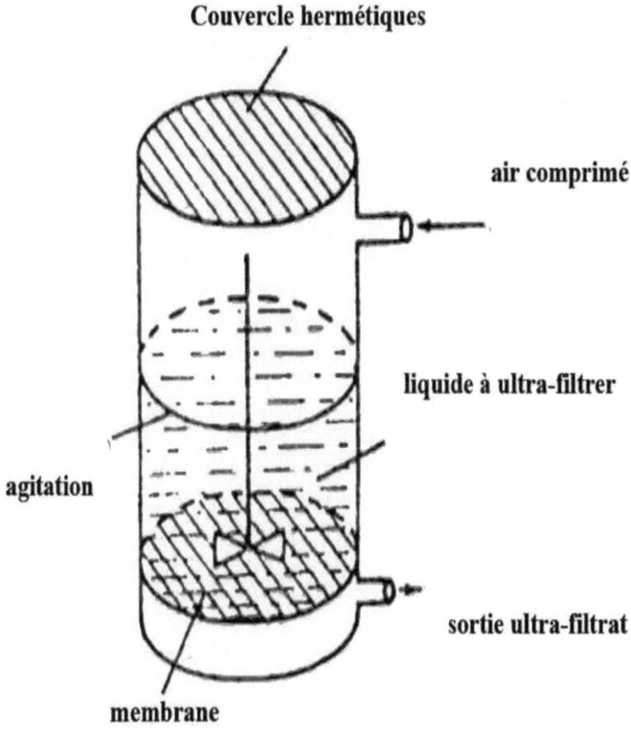

Figure 19: Ultrafiltration cell

Table 02: Ultrafiltration membranes

Manufacturers	Membranes	Constitution	Water permeability (100 psi pressure) in $ml.cm^{-2}. min^{-1}$	Molecular weight Substance- withheld at 20-100	Molecule used for measurement
Unio carbide	Cellophane gel	Cellulosic	0,004	10.000	Cytochrome c
Gelman	PEM Membrane	Cellulosic	0,02	40.000	Ovalbumin
Schleicher-Schuell	P. Membrane	Cellulosic	0,08	60.000	Albumin
General atomics	CA-Type B	Cellulose acetate	0,007	600	Raffinose
	CA-Type C		0,003	350	Sucrose
Amicon Corp	Diaflo UM-OS	Complex polyolectrolyte	0,05	350	Sucrose
	Diaflo UM-2		0,1	600	Raffinose
	Diaflo UM-10		0,3	10.000	Dextran-10
	Diaflo PM-10	Aromatic polymers	0,5	10.000	Cytochrome c
	Diaflo PM-30		0,07	30.000	Ovalbumin
	Diaflo XM-50	Olefins Substituted	0,07	50.000	Albumin
	Diaflo XM- 100A		0,9	100.000	Globulin 7 S
	Diaflo XM-300		1,1	300.000	Apoferritin
Abcor Corp	HFA-100	Cellulosic	0,07	10.000	Dextran-10
	HFA-200		0,40	20.000	Dextran-20
	HFA-300		1,40	70.000	Albumin
Millipore Corp	PSAC	Cellulosic	1,2	1.000	Bromo cresol green
	PSED		0,75	25.000	a-Chymotryp sine
	PSDM		1,0	40.000	Ovalbumin

2.2.2.1. Ultrafiltration on capillary membrane filters

Capillary membranes or hollow fibers have a wall thickness of 2.5 um, whereas ultrafiltration membranes have a thickness of 250 to 500 um. A very large number of these fibers are attached to a support and the assembly comprises (Figure 20) a double circuit: one concerning the internal volume of the fibers (or internal circuit), the other corresponding to the volume outside the fibers (external circuit).

There are four types of fibre:

HFU (hollow fiber ultrafiltration): exclusion limit: for substances with a molecular weight greater than 30,000; HFD (hollow fiber dialysis): exclusion limit: 5,000; HFO (hollow fiber osmosis): exclusion limit: 200; HFG (hollow fiber for gases).

Four setups are available: macrotube (Figure 20), beaker (Figure 21); micro beaker; micro T-tube (Figure 22).

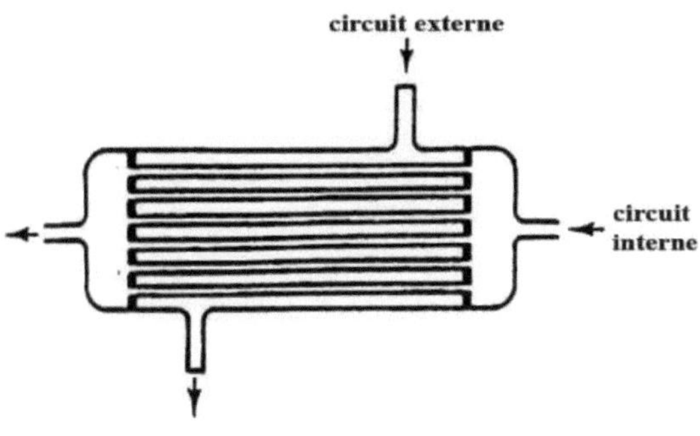

Figure 20: Diagram of capillary fibre assemblies (macrotube)

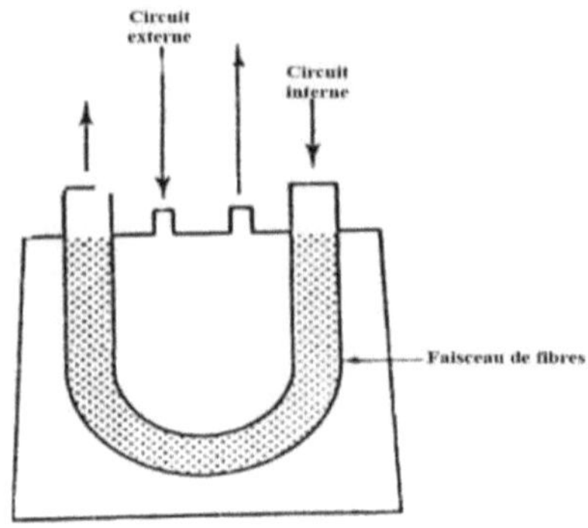

Figure 21: Beaker with capillary fibres.

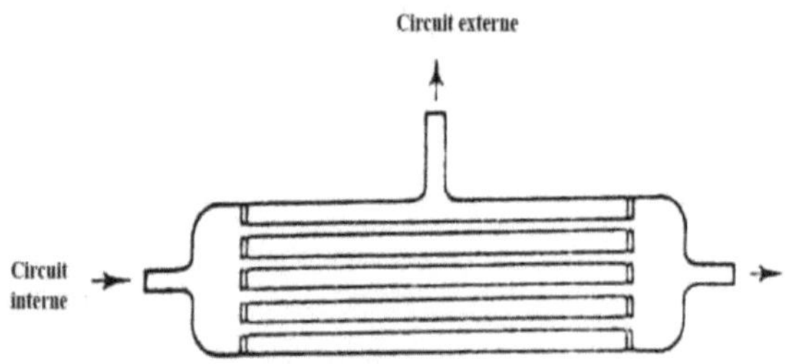

Figure 22: T-tube

A number of precautions must be taken when using these assemblies:

- Rinse with water at 20-30°C before use to remove the glycerine coating on the fibres;

- The liquids must not contain particles larger than 10 um, hence the need for pre-filtration on a Millipore filter; their temperature can vary from 0 to 60°C, their viscosity must be low, their pH between 3 and 10;

- The pressure inside the membranes must be less than 1 kg/cm^2 and 0,5 kg/cm^2 on the outside;

- Organic solvents destroy the fibres;

- After use, the assembly must be rinsed thoroughly and the membranes kept moist in a bacteriostatic solution;

- To check the tightness of the circuit, the pressure must be applied to the inlet of the device by closing the outlet, the whole device being immersed in water: the presence of air bubbles indicates the existence of a leak.

2.3. Applications

2.3.1. Desalting of biological samples

The liquid to be desalinated may flow into the inner compartment and the water into the outer compartment (Figure 23 A), or the liquid to be desalinated may be placed in the outer compartment and the water will then flow into the inner compartment (Figure 23 B). This desalting may be accompanied by a concentration of the sample if less water enters the internal circuit than is removed. Figure 23 C shows a set-up allowing a concentration of 1 ml/min.

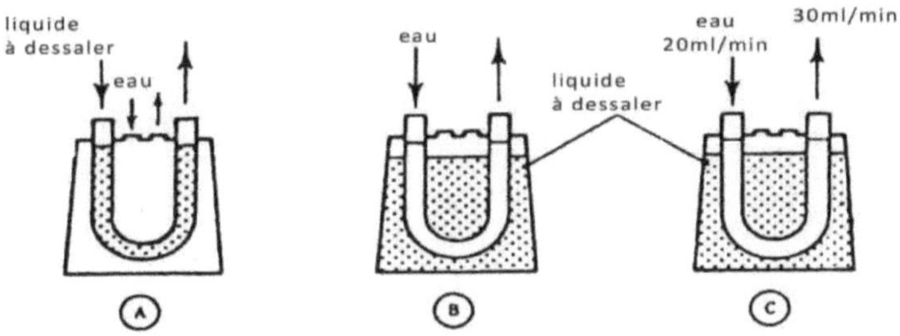

Figure 23: Desalting of biological fluids

2.3.2. Concentration of biological fluids

Low molecular weight products can be concentrated by osmosis using a 150g/l anhydrous CaCl2 solution or by dialysis against a 200g/l polyethylene glycol (or P.E.G.) solution. The liquid to be concentrated is placed in the internal (Figure 24 A) or external (Figure 24 B) circuit. The fibres used are of type 0 or D.

High molecular weight products are concentrated by establishing a pressure differential between the two compartments (by passing the solution under pressure or by creating a partial vacuum in the outer compartment) (Figure 24 C and D).

Low volume samples can be concentrated using the setups shown in Figure 24 E and 24 F. These can be used to continuously concentrate a chromatography column eluate.

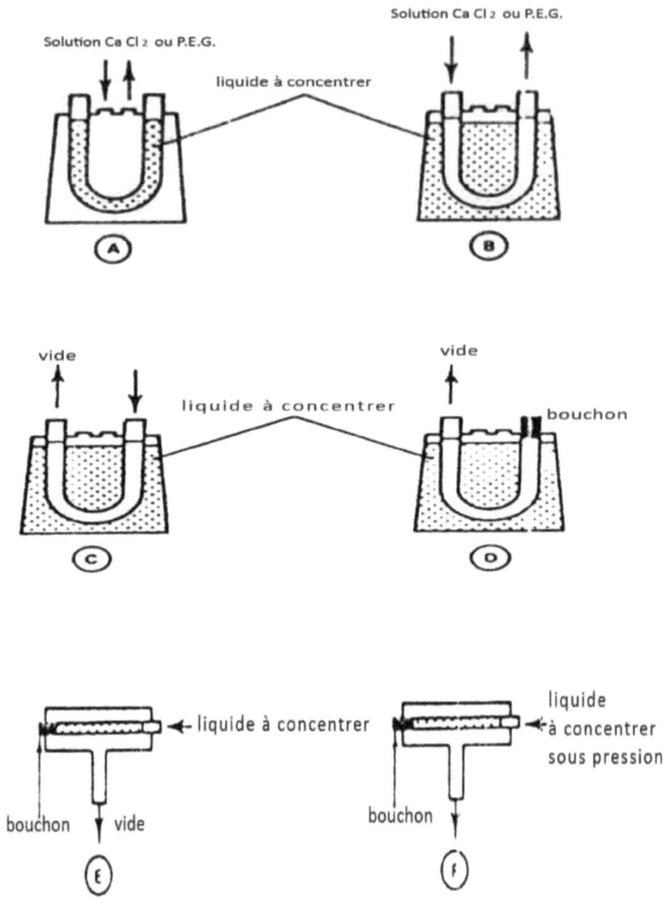

Figure 24: Concentration of biological fluids

3. CENTRIFUGATION, ULTRACENTRIFUGATION

Centrifugation is essentially a means of separating particles from a solution. In biology, particles are often cells, organelles (mitochondria, chloroplasts, etc.) or large molecules (proteins, nucleic acids, viruses, etc.).

Centrifugal separation is an accelerated variant of sedimentation where the natural acceleration of gravity ($g = 9.81\ ms^{-1}$) is replaced by a centrifugal force generated by the rotor of a high-speed centrifuge.

There are two types of centrifugation: one called preparative which allows the isolation of specific particles, the other is called analytical allowing the measurement of physical properties of the particles.

3.1 Basic theory of sedimentation

In the absence of agitation, particles dispersed in a fluid are subject, on the one hand, to the forces of gravity which tend to make them settle at the bottom of the container and, on the other hand, to Archimedes' buoyancy which tends to make them rise to the surface: the movement of the particles therefore depends on the relative intensity of these forces. If they are equal, the particles float, if gravity prevails, the particles sediment, and this all the faster as their difference in density with the dispersion medium is greater.

By rotating a rotor in a centrifuge, a centrifugal force is applied to each particle in the sample and each particle settles at a rate proportional to the centrifugal force applied to it. The viscosity of the solution containing the sample and the physical properties of the particle also affect the settling speed of each individual particle.

At a given centrifugal force and liquid viscosity, the sedimentation rate of the particle is proportional to the size (molecular weight) and the difference between its density and the solution.

During a centrifugation which tends to make them sediment and, consequently, to establish a gradient of concentration, from the top to the bottom of the centrifugation tube.

To the diffusion which tends, on the contrary, to maintain them in uniform solution opposing the establishment of a concentration gradient.

In order to allow the particles to settle, centrifugation speeds should be high enough that the centrifugal force, f, overcomes the diffusion forces (Figure 25).

The sedimentation coefficient, s, of a substance is its sedimentation rate dr/dt relative to the gravity field ω^2 r :

$$S = (dr/dt).(1/\omega^2 r)$$

Where r = distance to the axis of rotation (cm) œ = angular velocity (radian. sec^{-1})

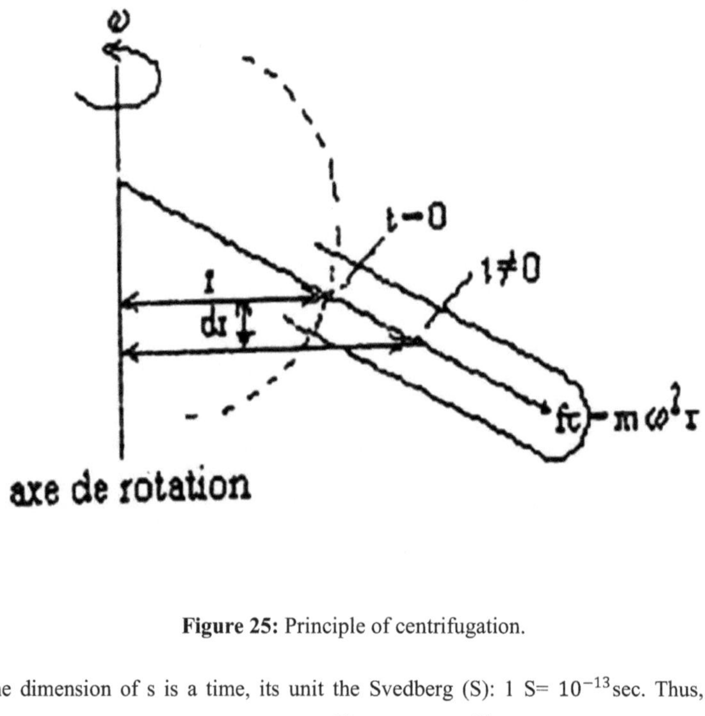

Figure 25: Principle of centrifugation.

The dimension of s is a time, its unit the Svedberg (S): 1 S= 10^{-13}sec. Thus, a particle whose sedimentation coefficient is equal to 10^{-12}sec = 10. 10^{-13} sec, has a value of 10 S.

Although the sedimentation coefficient increases with molecular weight, it is not proportional to it, since it is also influenced by the frictional resistances of the solvent and by the shape of the particle. However, under certain conditions, the molecular weight M of a protein can be calculated from the sedimentation coefficient using the Svedberg equation obtained by cancelling the centrifugal force by the opposing frictional force, which occurs when the sedimentation rate is constant:

- The centrifugal force to which a molecule of the dissolved substance is subjected is :

$F = M(1 - \vartheta\rho)\omega^2 r$

ϑ = specific partial volume of the substance. It corresponds to the increase in volume produced when 1g of dry solute is added to a large volume of solvent.

ρ = density of the solvent

- The frictional force is :

f' = F.dr/ dt

F= coefficient of friction, is equal to RT / D

R = perfect gas constant ($8.315\ 10^7$ ergs. mol^{-1}. $degré^{-1}$

T = absolute temperature (°K)

D= diffusion coefficient ($cm^2.\ s^{-1}$) in water at 20 °C, $D = D_{20,w}$

When f = f', we have:

$M(1 - \vartheta\rho)\omega^2 r = (RT/D)(dr/dt)$

Hence:

$M = RTs / D(1 - up)$ Svedberg equation

This equation shows that the mass of a particle is proportional to its sedimentation coefficient is inversely proportional to its diffusion coefficient. For example, a particle with a small molecular weight should have a small sedimentation coefficient and should show a tendency to diffuse rapidly in solutions of low viscosity.

3.2. Instrumentation

If the speed of rotation is less than 104 rpm, it is a centrifuge. The ultracentrifuge can reach 100,000 rpm or more and produce a gravity field sufficient to sediment particles or molecules with a mass greater than 10,000 daltons.

Centrifuges and ultracentrifuges consist of the following components:

- An engine capable of turning up to several tens of thousands of revolutions per minute,

- A rotor, capable of withstanding such fast rotations (usually made of titanium for the rotors of ultracentrifuges). The rotor has slots, located symmetrically on either side of the axis, to receive tubes containing the biological solutions.

- A chamber in which the rotor is placed, which is cooled and under high vacuum in the case of ultracentrifuges to avoid excessive heating due to the friction of the rotor against the air and to facilitate temperature control,

- This enclosure is also shielded to avoid accidents in case of rotor explosion.

The rotors are of three types:

- Fixed angle for the simplest separations between pellet (cells, organelles, membranes, proteins) and supernatant (Figure 26 A). These rotors are mainly used for sequential separations at increasing rotation speeds,

- With movable buckets, used for finer separations, because the centrifugal force is exerted in the direction of the length of the bucket (Figure 26 B). This is because the bucket is movable about its suspension point on the rotor axis and is oriented during rotation so that the axis remains collinear with the field. Particles can then be separated, essentially by density, using density gradients,

- Analytical, used for highly accurate measurement of sedimentation coefficients.

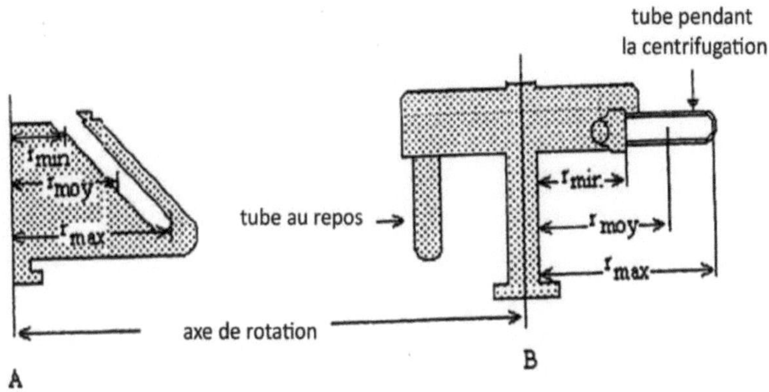

Figure 26: Angular rotor (A) and bucket rotor (B).

3.3. Methods of separation by preparative ultracentrifugation

3.3.1. Differential centrifugation

This is the most common method. It allows the isolation of intracellular structures, from an initial homogenate, by proceeding with different centrifugations at increasing speeds.

The centrifuge tube is initially filled with a uniform mixture of a sample solution. After centrifugation, initially at low acceleration (e.g. 1000g, 10 minutes), two fractions are separated: a pellet containing the denser elements that settle at the bottom of the centrifugation tubes and a supernatant solution of non-sedimented particles (Figure 27).

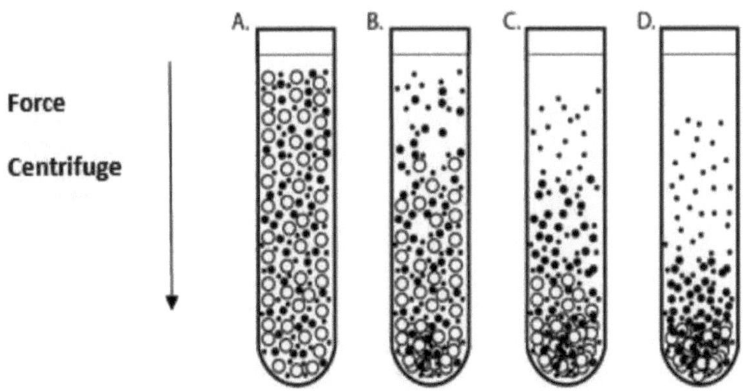

Figure 27: Differential centrifugation.

The two fractions are recovered by decanting the supernatant. The supernatant can be recentrifuged at higher speeds (e.g. 20,000g, 20 minutes) to obtain additional purification with the formation of a new pellet and a new supernatant (Figure 28). The pellet can be recentrifuged by constantly increasing the speed of rotation (e.g. 80 000g, 1 to 2 hours then 150 000g, 3 to 5 hours, etc.), after resuspension in a small volume of appropriate solvent. This method is used for fractionation of cellular components. The main steps required for complete fractionation of the cells are summarized in Figure 28. However, this ultracentrifugation technique does not allow the separation of all intracellular organelles. Thus, in most cases, the mitochondrial and lysosomal fractions are mixed and cannot be completely separated.

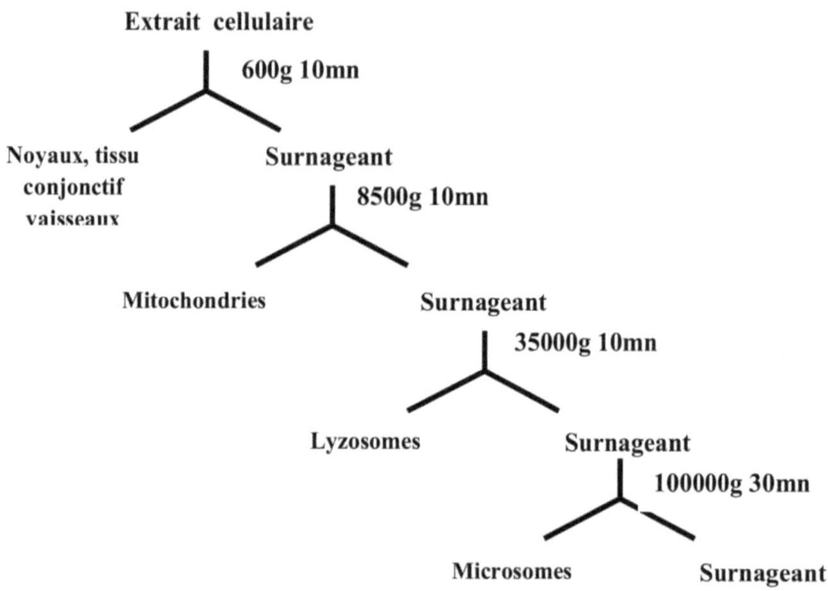

Figure 28: Cell fractionation by differential centrifugation.

3.3.2. Density gradient centrifugation

This technique is more complicated than the previous one but has additional advantages. It requires the establishment of a density gradient in the centrifuge tube. The density gradient fluid consists of a low molecular weight solute in a solvent containing the suspended sample particles. The density gradient used here is an auxiliary element: it simply serves to stabilize the separated products in zones by avoiding convection currents.

There are two methods of density gradient centrifugation:

3.3.2.1. Zone or zonal centrifugation

In this technique, a small volume of suspension containing the products to be analyzed is deposited on a preformed gradient column.

In most common techniques, a continuous density gradient (e.g. 5-45%) of sucrose (e.g. 5-45%) is first prepared in a centrifuge tube in

The system mixes a concentrated solution of sucrose with water in a decreasing ratio as the tube is filled, so that the density of the medium is greatest at the lower end of the tube (Figure 29). A small volume of the mixture to be separated is deposited at the upper end of the gradient. The whole mixture is centrifuged at speeds and for times that depend on the purpose of the centrifugation. Centrifuging the tube at high speed in a horizontal bucket rotor causes each type of particle to settle at its own speed, which is determined mainly by the weight of the particle, but also by its density and shape, resulting in separate bands or zones.

Usually, centrifugation is stopped when equilibrium is reached. The position of the bands can be located optically or by carefully collecting the contents of the tube by piercing a small hole in its lower end; successive samples are collected in a series of small tubes to be analysed in the order of collection. By determining the protein concentration for each fraction by measuring the optical density at 280 nm, the distribution of the proteins in the tube is determined, allowing them to be classified by mass.

The plastic tube can also be frozen and cut into thin layers for further analysis.

This technique is mainly used to separate DNA, RNA molecules or ribonucleoprotein structures according to their size.

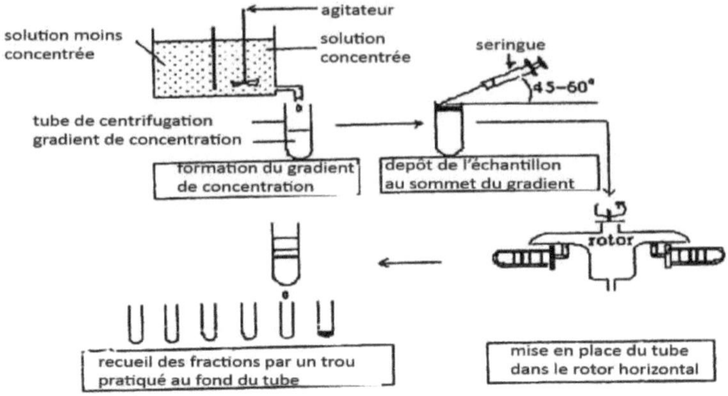

Figure 29: Conducting a zonal centrifugation.

3.3.2.2. Isopycnic centrifugation

This type of centrifugation uses a very dense solution of a salt such as 6M cesium chloride with which the sample to be separated is mixed. During centrifugation, the cesium chloride sedimentates, forming a stable concentration gradient between the top and bottom of the tube (Figure 30). This concentration change is associated with a density change or density gradient. This density gradient reaches a state of equilibrium after a certain centrifugation time, which depends on the centrifugation speed, the height of the liquid column, the nature of the salt and the temperature.

If a particle is initially mixed with the salt solution and the density of the particle surrounds that of the salt solution, the particle will form a more or less narrow band in the centrifuge tube at the position where its density is equal to that of the salt solution. The particles are then said to be isopycnic with respect to the medium. There are tables providing the densities of many particles (membrane vesicles, organelles, etc.) which allow the separation of the different particles from each other by choosing the appropriate gradient.

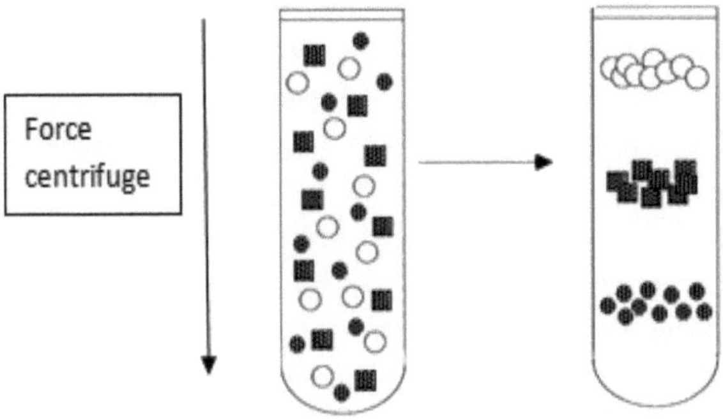

Figure 30: Isopycnic centrifugation.

When two species of macromolecules with different bulk densities are centrifuged, they occupy different positions in the centrifuge tube, and a separation of macromolecules is achieved based on their different bulk densities and not on their sedimentation coefficients. Even in a very strong gravity field, the particles cannot sediment beyond the region of the gradient of the same density as themselves.

Density gradients are established using compounds that are highly soluble in water and capable of changing the density of the medium by sedimentation. The compound most used in the formation of a density gradient for equilibrium sedimentation is an aqueous solution of chloride (CsCl). A gradient containing more Cs+ ions (and more Cl- which moves with the Cs+ to neutralize the charges) is established at the bottom of the tube than at the top. In a typical centrifugation, the centrifugation medium is about 0.02 mg/ml denser than in a CsCl solution, are about 1.3 to 1.6 to 1.7 and 1.75 to 1.8 g/ml respectively, and therefore easily separable.

In a CsCl solution, the ions bind to proteins and nucleic acids; the Cs+ ion binds to the phosphate groups of DNA, and in RNA, to the phosphate groups as well as to the hydroxyl functions of the riboses, which increases the density of RNA more than that of DNA.

To separate proteins or organelles, concentration gradients of sucrose or dextran (polymerized sugar) are often used.

This equilibrium centrifugation has a very high resolution and is particularly useful for distinguishing macromolecules where the usual ^{12}C or ^{14}N is substituted by the heavy isotopes ^{13}C or ^{15}N. Thus, when cells are grown with heavy isotopes, it is possible to physically separate the molecules made by the cell before the addition of these heavy isotopes from those made afterwards. A perfect illustration of this application was Stahl's experiment showing the semi-conservative replication of DNA.

3.4. Separation by analytical ultracentrifugation

An analytical centrifuge (Figure 31) has an observation and measurement system that allows the migration of particles in a homogeneous solution to be followed during the experiment, thus allowing the sedimentation coefficient (s) to be determined, using an optical measurement of the concentration (absorbance) or a variation in concentration (refractive index difference).

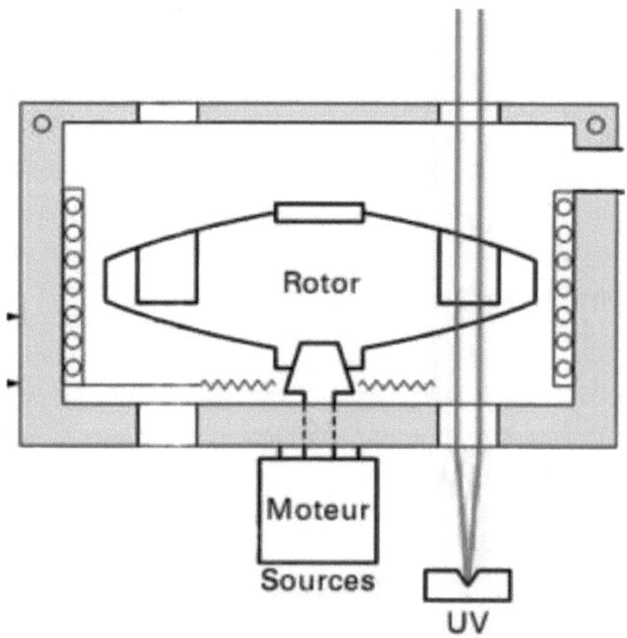

Figure 31: Analytical ultracentrifuge.

3.5. Control of the centrifugation result

Whatever the technical variant used for the preparation of cell fractions, it must always be ensured that the fraction collected contains only the organelle or particles to be studied and that there is no possible contamination by another cell fraction in this preparation. This control can be done either by morphological criteria (electron microscopy) or by biochemical criteria (search for marker enzymes).

Thus, acid phosphatases and p-glucuronidase characterize lysosomal fractions; catalase, peroxisomes. Bone transferases are specific for golgitic vesicles and 5'-nucleotidase is the most specific enzyme for the cytoplasmic membrane. Finally, for mitochondria, the outer membrane may be marked by the presence of monoamine oxidase while the inner membrane is marked by the presence of succino-deshydrogenase.

Moreover, it is advisable to choose the tissue to be used according to the organelle to be studied: liver tissue is particularly suitable for the study of mitochondria, whereas red blood cells are the material of choice for any research on plasma membranes.

3.6. Some aspects of the practical implementation of centrifugations

The density gradients used in isopycnic centrifugation are equilibrium gradients; they are established during centrifugation.

In general, the macromolecules to be centrifuged are initially mixed with the saline solution, the density of which has been suitably chosen.

If it is proposed to prepare v ml of a solution whose density is p, the mass $_{mo}$ of salt to be used is given by :

$$m_0 = v.p.F$$

where F is the weight fraction of the salt in a given aqueous solution and the mass m of buffer, or macromolecular solution or water to be added is :

$$M = v.\rho - m_0$$

Thus, to obtain 3ml of CsCl at density p =1.760 (F = 0.589), we weighed :

m_0 = 3 . 1,760 - 3,110 = 2,170 ml of buffer.

It is a good idea to check that the final solution has the desired density. This can be done either by pycnometry or by refractometry. In the second case, the relationship between density and refractive index of concentrated salt solutions is used.

Example: Case of CsCl at 25°C

$\rho^{25} = 10{,}2402\, \eta^{25} - 12.6483$ for densities between 1.00 and 1.38

$\rho^{25} = 10{,}681\, \eta^{25} - 13.4974$ for densities above 1.37.

At the end of the centrifugation, the macromolecules form narrow bands in the centrifuge tube and the position, if any, can be determined. For this purpose, the contents of the centrifuge tube are collected in the same way as during centrifugation of the zone....

Bibliographic references:

- Biochemistry. Donald Voet, Judith G. Voet. 4 th edition. Wiley .2016. Page 141.

- Centrifugation: Essential Data. David Rickwood, T. Ford, J. Steensgaard. Wiley, June 14, 1994 - 114 pages.

- Instrumental analysis. A manual for biologists. Marouf abdelrrazak. 2nd edition 2005. DAR EL GHARB.

- Apparatus and methods in biochemistry. P Kamoun ,Flammarion, Médecine-Sciences, Paris (1977).

Dialysis. Jean PASTOR, Anne-Marie PAULI. Professors at the University of Aix-Marseille II, Faculty of Pharmacy: Laboratory of Analytical Chemistry. Engineering techniques.

https://www.techniques-ingenieur.fr/

- Ultracentrifugation. Lucette BARDET ; Professor at the Faculty of Pharmacy of Montpellier. Editions : Techniques de l'ingénieur. https://www.techniques-ingenieur.fr/

- Decantation - Filtration. Jean HACHE, Scientific and Development Director of ATTOBIO. Editions: Techniques de l'ingénieur. https://www.techniques-ingenieur.fr/

SOME WEBSITES ON THE INTERNET

https://books.google.dz/books?isbn=052165873X

https://books.google.dz/books?isbn=0306466546

https://books.google.dz/books?isbn=1428900136

https://books.google.dz/books?isbn=3540678689

MIX
Papier aus verantwortungsvollen Quellen
Paper from responsible sources
FSC® C105338

Printed by Books on Demand GmbH, Norderstedt / Germany